PILOT FLIGHT LOGBOOK

Logbook Number _____

Period _____ ~ _____

PILOT'S NAME
PERMANENT MAILING ADDRESS
CHANGE OF ADDRESS
E-MAIL / MOBILE

RECORD OF CERTIFICATES and RATINGS

CERTIFICATES			RATINGS		
TYPE	NUMBER	DATE OF ISSUE	CATEGORY, CLASS, OR TYPE	DATE OF ISSUE	TOTAL TIME

AIRCRAFT RECORD BY MODEL													FLIGHT PROFICIENCY				
MODEL _____ YEAR _____				MODEL _____ YEAR _____				MODEL _____ YEAR _____					RECURRENT CHECK	PIC PROFICIENCY CHECK	LIP RECURRENT TRAINING	CFI RENEWAL	
	PIC	SIC	IP		PIC	SIC	IP		PIC	SIC	IP						
JAN.																	
FEB.																	
MAR.																	
APR.																	
MAY.																	
JUN.																	
TOTALS 1ST 6 MO.																	
JUL.																	
AUG.																	
SEP.																	
OCT.																	
NOV.																	
DEC.																	
YEAR TOTALS																	
TOTALS PREV. YEARS																	
TOTALS TO DATE																	

20 _____

DATE (MON/DAY)	AIRCRAFT TYPE	AIRCRAFT IDENT	ROUTE OF FLIGHT		FLIGHT NUMBER	TAKE OFF & LANDING			CONDITION OF FLIGHT		APP		FLIGHT SIMULATOR
			FROM	TO		DAY	NIGHT	AUTO LAND	NIGHT	ACTUAL INSTRUMENT	R/W NO.	TYPE	
						/	/						
						/	/						
						/	/						
						/	/						
						/	/						
						/	/						
						/	/						
						/	/						
						/	/						
						/	/						
						/	/						
						/	/						
THIS RECORD IS CERTIFIED TRUE AND CORRECT PILOT'S SIGNATURE _____			PAGE TOTAL			/	/						
			PREVIOUS TOTAL			/	/						
			NEW TOTAL			/	/						

TYPE OF PILOTING TIME						REMARKS AND ENDORSEMENTS
PIC	SIC	AS FLIGHT INSTRUCTOR	DUAL RECEIVED			

20 _____

| DATE (MON/DAY) | AIRCRAFT TYPE | AIRCRAFT IDENT | ROUTE OF FLIGHT | | FLIGHT NUMBER | TAKE OFF & LANDING | | | CONDITION OF FLIGHT | | APP | | FLIGHT SIMULATOR |
			FROM	TO		DAY	NIGHT	AUTO LAND	NIGHT	ACTUAL INSTRUMENT	R/W NO.	TYPE	
						/	/						
						/	/						
						/	/						
						/	/						
						/	/						
						/	/						
						/	/						
						/	/						
						/	/						
						/	/						
						/	/						
THIS RECORD IS CERTIFIED TRUE AND CORRECT PILOT'S SIGNATURE _____			PAGE TOTAL			/	/						
			PREVIOUS TOTAL			/	/						
			NEW TOTAL			/	/						

TYPE OF PILOTING TIME							REMARKS AND ENDORSEMENTS
PIC	SIC	AS FLIGHT INSTRUCTOR	DUAL RECEIVED				

20 _____

DATE (MON/DAY)	AIRCRAFT TYPE	AIRCRAFT IDENT	ROUTE OF FLIGHT		FLIGHT NUMBER	TAKE OFF & LANDING			CONDITION OF FLIGHT				FLIGHT SIMULATOR
			FROM	TO		DAY	NIGHT	AUTO LAND	NIGHT	ACTUAL INSTRUMENT	APP R/W NO.	TYPE	
						/	/						
						/	/						
						/	/						
						/	/						
						/	/						
						/	/						
						/	/						
						/	/						
						/	/						
						/	/						
						/	/						
THIS RECORD IS CERTIFIED TRUE AND CORRECT PILOT'S SIGNATURE _____			PAGE TOTAL			/	/						
			PREVIOUS TOTAL			/	/						
			NEW TOTAL			/	/						

		TYPE OF PILOTING TIME					REMARKS AND ENDORSEMENTS
PIC	SIC	AS FLIGHT INSTRUCTOR	DUAL RECEIVED				

20 _____

DATE (MON/DAY)	AIRCRAFT TYPE	AIRCRAFT IDENT	ROUTE OF FLIGHT		FLIGHT NUMBER	TAKE OFF & LANDING			CONDITION OF FLIGHT		APP		FLIGHT SIMULATOR
			FROM	TO		DAY	NIGHT	AUTO LAND	NIGHT	ACTUAL INSTRUMENT	R/W NO.	TYPE	
						/	/						
						/	/						
						/	/						
						/	/						
						/	/						
						/	/						
						/	/						
						/	/						
						/	/						
						/	/						
						/	/						
THIS RECORD IS CERTIFIED TRUE AND CORRECT PILOT'S SIGNATURE _____				PAGE TOTAL		/	/						
^				PREVIOUS TOTAL		/	/						
^				NEW TOTAL		/	/						

TYPE OF PILOTING TIME							REMARKS AND ENDORSEMENTS
PIC	SIC	AS FLIGHT INSTRUCTOR	DUAL RECEIVED				

20 _____

DATE (MON/DAY)	AIRCRAFT TYPE	AIRCRAFT IDENT	ROUTE OF FLIGHT		FLIGHT NUMBER	TAKE OFF & LANDING			CONDITION OF FLIGHT		APP		FLIGHT SIMULATOR
			FROM	TO		DAY	NIGHT	AUTO LAND	NIGHT	ACTUAL INSTRUMENT	R/W NO.	TYPE	
						/	/						
						/	/						
						/	/						
						/	/						
						/	/						
						/	/						
						/	/						
						/	/						
						/	/						
						/	/						
						/	/						
THIS RECORD IS CERTIFIED TRUE AND CORRECT PILOT'S SIGNATURE _____			PAGE TOTAL			/	/						
			PREVIOUS TOTAL			/	/						
			NEW TOTAL			/	/						

PIC	SIC	AS FLIGHT INSTRUCTOR	DUAL RECEIVED			REMARKS AND ENDORSEMENTS

TYPE OF PILOTING TIME

20 _____

DATE (MON/DAY)	AIRCRAFT TYPE	AIRCRAFT IDENT	ROUTE OF FLIGHT		FLIGHT NUMBER	TAKE OFF & LANDING			CONDITION OF FLIGHT		APP		FLIGHT SIMULATOR
			FROM	TO		DAY	NIGHT	AUTO LAND	NIGHT	ACTUAL INSTRUMENT	R/W NO.	TYPE	
						/	/						
						/	/						
						/	/						
						/	/						
						/	/						
						/	/						
						/	/						
						/	/						
						/	/						
						/	/						
						/	/						
THIS RECORD IS CERTIFIED TRUE AND CORRECT PILOT'S SIGNATURE _____			PAGE TOTAL			/	/						
			PREVIOUS TOTAL			/	/						
			NEW TOTAL			/	/						

TYPE OF PILOTING TIME							REMARKS AND ENDORSEMENTS
PIC	SIC	AS FLIGHT INSTRUCTOR	DUAL RECEIVED				

20 _____

DATE (MON/DAY)	AIRCRAFT TYPE	AIRCRAFT IDENT	ROUTE OF FLIGHT		FLIGHT NUMBER	TAKE OFF & LANDING			CONDITION OF FLIGHT				FLIGHT SIMULATOR
											APP		
			FROM	TO		DAY	NIGHT	AUTO LAND	NIGHT	ACTUAL INSTRUMENT	R/W NO.	TYPE	
						/	/						
						/	/						
						/	/						
						/	/						
						/	/						
						/	/						
						/	/						
						/	/						
						/	/						
						/	/						
						/	/						
						/	/						
THIS RECORD IS CERTIFIED TRUE AND CORRECT PILOT'S SIGNATURE _____			PAGE TOTAL			/	/						
^^^			PREVIOUS TOTAL			/	/						
^^^			NEW TOTAL			/	/						

		TYPE OF PILOTING TIME					REMARKS AND ENDORSEMENTS
PIC	SIC	AS FLIGHT INSTRUCTOR	DUAL RECEIVED				

20 _____

DATE (MON/DAY)	AIRCRAFT TYPE	AIRCRAFT IDENT	ROUTE OF FLIGHT		FLIGHT NUMBER	TAKE OFF & LANDING			CONDITION OF FLIGHT		APP		FLIGHT SIMULATOR
			FROM	TO		DAY	NIGHT	AUTO LAND	NIGHT	ACTUAL INSTRUMENT	R/W NO.	TYPE	
						/	/						
						/	/						
						/	/						
						/	/						
						/	/						
						/	/						
						/	/						
						/	/						
						/	/						
						/	/						
						/	/						
						/	/						
THIS RECORD IS CERTIFIED TRUE AND CORRECT PILOT'S SIGNATURE _____			PAGE TOTAL			/	/						
			PREVIOUS TOTAL			/	/						
			NEW TOTAL			/	/						

		TYPE OF PILOTING TIME					REMARKS AND ENDORSEMENTS
PIC	SIC	AS FLIGHT INSTRUCTOR	DUAL RECEIVED				

20 _____

| DATE (MON/DAY) | AIRCRAFT TYPE | AIRCRAFT IDENT | ROUTE OF FLIGHT | | FLIGHT NUMBER | TAKE OFF & LANDING | | | CONDITION OF FLIGHT | | APP | | FLIGHT SIMULATOR |
			FROM	TO		DAY	NIGHT	AUTO LAND	NIGHT	ACTUAL INSTRUMENT	R/W NO.	TYPE	
						/	/						
						/	/						
						/	/						
						/	/						
						/	/						
						/	/						
						/	/						
						/	/						
						/	/						
						/	/						
						/	/						
THIS RECORD IS CERTIFIED TRUE AND CORRECT PILOT'S SIGNATURE _____			PAGE TOTAL			/	/						
			PREVIOUS TOTAL			/	/						
			NEW TOTAL			/	/						

TYPE OF PILOTING TIME						REMARKS AND ENDORSEMENTS
PIC	SIC	AS FLIGHT INSTRUCTOR	DUAL RECEIVED			

20 _____

DATE (MON/DAY)	AIRCRAFT TYPE	AIRCRAFT IDENT	ROUTE OF FLIGHT		FLIGHT NUMBER	TAKE OFF & LANDING			CONDITION OF FLIGHT				FLIGHT SIMULATOR
			FROM	TO		DAY	NIGHT	AUTO LAND	NIGHT	ACTUAL INSTRUMENT	APP R/W NO.	APP TYPE	
						/	/						
						/	/						
						/	/						
						/	/						
						/	/						
						/	/						
						/	/						
						/	/						
						/	/						
						/	/						
						/	/						
THIS RECORD IS CERTIFIED TRUE AND CORRECT PILOT'S SIGNATURE _____			PAGE TOTAL		/	/						
			PREVIOUS TOTAL		/	/						
			NEW TOTAL		/	/						

TYPE OF PILOTING TIME						REMARKS AND ENDORSEMENTS
PIC	SIC	AS FLIGHT INSTRUCTOR	DUAL RECEIVED			

20 _____

DATE (MON/DAY)	AIRCRAFT TYPE	AIRCRAFT IDENT	ROUTE OF FLIGHT		FLIGHT NUMBER	TAKE OFF & LANDING			CONDITION OF FLIGHT				FLIGHT SIMULATOR
			FROM	TO		DAY	NIGHT	AUTO LAND	NIGHT	ACTUAL INSTRUMENT	APP		
											R/W NO.	TYPE	
						/	/						
						/	/						
						/	/						
						/	/						
						/	/						
						/	/						
						/	/						
						/	/						
						/	/						
						/	/						
						/	/						
THIS RECORD IS CERTIFIED TRUE AND CORRECT PILOT'S SIGNATURE _____				PAGE TOTAL		/	/						
^				PREVIOUS TOTAL		/	/						
^				NEW TOTAL		/	/						

	TYPE OF PILOTING TIME						REMARKS AND ENDORSEMENTS
PIC	SIC	AS FLIGHT INSTRUCTOR	DUAL RECEIVED				

20 _____

DATE (MON/DAY)	AIRCRAFT TYPE	AIRCRAFT IDENT	ROUTE OF FLIGHT		FLIGHT NUMBER	TAKE OFF & LANDING			CONDITION OF FLIGHT		APP		FLIGHT SIMULATOR
			FROM	TO		DAY	NIGHT	AUTO LAND	NIGHT	ACTUAL INSTRUMENT	R/W NO.	TYPE	
						/	/						
						/	/						
						/	/						
						/	/						
						/	/						
						/	/						
						/	/						
						/	/						
						/	/						
						/	/						
						/	/						
THIS RECORD IS CERTIFIED TRUE AND CORRECT PILOT'S SIGNATURE _____			PAGE TOTAL			/	/						
			PREVIOUS TOTAL			/	/						
			NEW TOTAL			/	/						

TYPE OF PILOTING TIME						REMARKS AND ENDORSEMENTS
PIC	SIC	AS FLIGHT INSTRUCTOR	DUAL RECEIVED			

20 _____

DATE (MON/DAY)	AIRCRAFT TYPE	AIRCRAFT IDENT	ROUTE OF FLIGHT		FLIGHT NUMBER	TAKE OFF & LANDING			CONDITION OF FLIGHT		APP		FLIGHT SIMULATOR
			FROM	TO		DAY	NIGHT	AUTO LAND	NIGHT	ACTUAL INSTRUMENT	R/W NO.	TYPE	
						/	/						
						/	/						
						/	/						
						/	/						
						/	/						
						/	/						
						/	/						
						/	/						
						/	/						
						/	/						
						/	/						
THIS RECORD IS CERTIFIED TRUE AND CORRECT PILOT'S SIGNATURE _____			PAGE TOTAL			/	/						
			PREVIOUS TOTAL			/	/						
			NEW TOTAL			/	/						

TYPE OF PILOTING TIME							REMARKS AND ENDORSEMENTS
PIC	SIC	AS FLIGHT INSTRUCTOR	DUAL RECEIVED			

20 _____

DATE (MON/DAY)	AIRCRAFT TYPE	AIRCRAFT IDENT	ROUTE OF FLIGHT		FLIGHT NUMBER	TAKE OFF & LANDING			CONDITION OF FLIGHT				FLIGHT SIMULATOR
			FROM	TO		DAY	NIGHT	AUTO LAND	NIGHT	ACTUAL INSTRUMENT	APP R/W NO.	TYPE	
						/	/						
						/	/						
						/	/						
						/	/						
						/	/						
						/	/						
						/	/						
						/	/						
						/	/						
						/	/						
						/	/						
THIS RECORD IS CERTIFIED TRUE AND CORRECT PILOT'S SIGNATURE _____			PAGE TOTAL			/	/						
			PREVIOUS TOTAL			/	/						
			NEW TOTAL			/	/						

TYPE OF PILOTING TIME						REMARKS AND ENDORSEMENTS
PIC	SIC	AS FLIGHT INSTRUCTOR	DUAL RECEIVED			

20 _____

DATE (MON/DAY)	AIRCRAFT TYPE	AIRCRAFT IDENT	ROUTE OF FLIGHT		FLIGHT NUMBER	TAKE OFF & LANDING			CONDITION OF FLIGHT		APP		FLIGHT SIMULATOR
			FROM	TO		DAY	NIGHT	AUTO LAND	NIGHT	ACTUAL INSTRUMENT	R/W NO.	TYPE	
						/	/						
						/	/						
						/	/						
						/	/						
						/	/						
						/	/						
						/	/						
						/	/						
						/	/						
						/	/						
						/	/						
THIS RECORD IS CERTIFIED TRUE AND CORRECT PILOT'S SIGNATURE _____			PAGE TOTAL			/	/						
			PREVIOUS TOTAL			/	/						
			NEW TOTAL			/	/						

TYPE OF PILOTING TIME						REMARKS AND ENDORSEMENTS
PIC	SIC	AS FLIGHT INSTRUCTOR	DUAL RECEIVED			

20 _____

DATE (MON/DAY)	AIRCRAFT TYPE	AIRCRAFT IDENT	ROUTE OF FLIGHT		FLIGHT NUMBER	TAKE OFF & LANDING			CONDITION OF FLIGHT				FLIGHT SIMULATOR
			FROM	TO		DAY	NIGHT	AUTO LAND	NIGHT	ACTUAL INSTRUMENT	APP		
											R/W NO.	TYPE	
						/	/						
						/	/						
						/	/						
						/	/						
						/	/						
						/	/						
						/	/						
						/	/						
						/	/						
						/	/						
						/	/						
THIS RECORD IS CERTIFIED TRUE AND CORRECT PILOT'S SIGNATURE _____			PAGE TOTAL			/	/						
			PREVIOUS TOTAL			/	/						
			NEW TOTAL			/	/						

		TYPE OF PILOTING TIME					REMARKS AND ENDORSEMENTS
PIC	SIC	AS FLIGHT INSTRUCTOR	DUAL RECEIVED				

20 _____

DATE (MON/DAY)	AIRCRAFT TYPE	AIRCRAFT IDENT	ROUTE OF FLIGHT		FLIGHT NUMBER	TAKE OFF & LANDING			CONDITION OF FLIGHT				FLIGHT SIMULATOR
			FROM	TO		DAY	NIGHT	AUTO LAND	NIGHT	ACTUAL INSTRUMENT	APP R/W NO.	TYPE	
						/	/						
						/	/						
						/	/						
						/	/						
						/	/						
						/	/						
						/	/						
						/	/						
						/	/						
						/	/						
						/	/						

THIS RECORD IS CERTIFIED TRUE AND CORRECT
PILOT'S SIGNATURE _____

	PAGE TOTAL	/	/					
	PREVIOUS TOTAL	/	/					
	NEW TOTAL	/	/					

TYPE OF PILOTING TIME						REMARKS AND ENDORSEMENTS
PIC	SIC	AS FLIGHT INSTRUCTOR	DUAL RECEIVED			

20 _____

DATE (MON/DAY)	AIRCRAFT TYPE	AIRCRAFT IDENT	ROUTE OF FLIGHT		FLIGHT NUMBER	TAKE OFF & LANDING			CONDITION OF FLIGHT				FLIGHT SIMULATOR
			FROM	TO		DAY	NIGHT	AUTO LAND	NIGHT	ACTUAL INSTRUMENT	APP R/W NO.	TYPE	
						/	/						
						/	/						
						/	/						
						/	/						
						/	/						
						/	/						
						/	/						
						/	/						
						/	/						
						/	/						
						/	/						

THIS RECORD IS CERTIFIED TRUE AND CORRECT PILOT'S SIGNATURE _____				
	PAGE TOTAL	/	/	
	PREVIOUS TOTAL	/	/	
	NEW TOTAL	/	/	

TYPE OF PILOTING TIME						REMARKS AND ENDORSEMENTS
PIC	SIC	AS FLIGHT INSTRUCTOR	DUAL RECEIVED			

20 _____

DATE (MON/DAY)	AIRCRAFT TYPE	AIRCRAFT IDENT	ROUTE OF FLIGHT		FLIGHT NUMBER	TAKE OFF & LANDING			CONDITION OF FLIGHT		APP		FLIGHT SIMULATOR
			FROM	TO		DAY	NIGHT	AUTO LAND	NIGHT	ACTUAL INSTRUMENT	R/W NO.	TYPE	
						/	/						
						/	/						
						/	/						
						/	/						
						/	/						
						/	/						
						/	/						
						/	/						
						/	/						
						/	/						
						/	/						
THIS RECORD IS CERTIFIED TRUE AND CORRECT PILOT'S SIGNATURE _____			PAGE TOTAL			/	/						
			PREVIOUS TOTAL			/	/						
			NEW TOTAL			/	/						

		TYPE OF PILOTING TIME				REMARKS AND ENDORSEMENTS
PIC	SIC	AS FLIGHT INSTRUCTOR	DUAL RECEIVED			

20 _____

DATE (MON/DAY)	AIRCRAFT TYPE	AIRCRAFT IDENT	ROUTE OF FLIGHT		FLIGHT NUMBER	TAKE OFF & LANDING			CONDITION OF FLIGHT		APP		FLIGHT SIMULATOR
			FROM	TO		DAY	NIGHT	AUTO LAND	NIGHT	ACTUAL INSTRUMENT	R/W NO.	TYPE	
						/	/						
						/	/						
						/	/						
						/	/						
						/	/						
						/	/						
						/	/						
						/	/						
						/	/						
						/	/						
						/	/						
THIS RECORD IS CERTIFIED TRUE AND CORRECT PILOT'S SIGNATURE _____			PAGE TOTAL			/	/						
			PREVIOUS TOTAL			/	/						
			NEW TOTAL			/	/						

TYPE OF PILOTING TIME						REMARKS AND ENDORSEMENTS
PIC	SIC	AS FLIGHT INSTRUCTOR	DUAL RECEIVED			

20 _____

DATE (MON/DAY)	AIRCRAFT TYPE	AIRCRAFT IDENT	ROUTE OF FLIGHT		FLIGHT NUMBER	TAKE OFF & LANDING			CONDITION OF FLIGHT				FLIGHT SIMULATOR
			FROM	TO		DAY	NIGHT	AUTO LAND	NIGHT	ACTUAL INSTRUMENT	APP		
											R/W NO.	TYPE	
						/	/						
						/	/						
						/	/						
						/	/						
						/	/						
						/	/						
						/	/						
						/	/						
						/	/						
						/	/						
						/	/						
THIS RECORD IS CERTIFIED TRUE AND CORRECT PILOT'S SIGNATURE _____			PAGE TOTAL		/	/						
			PREVIOUS TOTAL		/	/						
			NEW TOTAL		/	/						

TYPE OF PILOTING TIME						REMARKS AND ENDORSEMENTS
PIC	SIC	AS FLIGHT INSTRUCTOR	DUAL RECEIVED			

20 _____

DATE (MON/DAY)	AIRCRAFT TYPE	AIRCRAFT IDENT	ROUTE OF FLIGHT		FLIGHT NUMBER	TAKE OFF & LANDING			CONDITION OF FLIGHT				FLIGHT SIMULATOR
			FROM	TO		DAY	NIGHT	AUTO LAND	NIGHT	ACTUAL INSTRUMENT	APP R/W NO.	TYPE	
						/	/						
						/	/						
						/	/						
						/	/						
						/	/						
						/	/						
						/	/						
						/	/						
						/	/						
						/	/						
						/	/						
THIS RECORD IS CERTIFIED TRUE AND CORRECT PILOT'S SIGNATURE _____			PAGE TOTAL			/	/						
			PREVIOUS TOTAL			/	/						
			NEW TOTAL			/	/						

TYPE OF PILOTING TIME						REMARKS AND ENDORSEMENTS
PIC	SIC	AS FLIGHT INSTRUCTOR	DUAL RECEIVED			

20 _____

DATE (MON/DAY)	AIRCRAFT TYPE	AIRCRAFT IDENT	ROUTE OF FLIGHT		FLIGHT NUMBER	TAKE OFF & LANDING			CONDITION OF FLIGHT				FLIGHT SIMULATOR
			FROM	TO		DAY	NIGHT	AUTO LAND	NIGHT	ACTUAL INSTRUMENT	APP		
											R/W NO.	TYPE	
						/	/						
						/	/						
						/	/						
						/	/						
						/	/						
						/	/						
						/	/						
						/	/						
						/	/						
						/	/						
						/	/						
THIS RECORD IS CERTIFIED TRUE AND CORRECT PILOT'S SIGNATURE _____			PAGE TOTAL		/	/						
			PREVIOUS TOTAL		/	/						
			NEW TOTAL		/	/						

		TYPE OF PILOTING TIME				REMARKS AND ENDORSEMENTS
PIC	SIC	AS FLIGHT INSTRUCTOR	DUAL RECEIVED			

20 _____

DATE (MON/DAY)	AIRCRAFT TYPE	AIRCRAFT IDENT	ROUTE OF FLIGHT		FLIGHT NUMBER	TAKE OFF & LANDING			CONDITION OF FLIGHT				FLIGHT SIMULATOR
			FROM	TO		DAY	NIGHT	AUTO LAND	NIGHT	ACTUAL INSTRUMENT	APP R/W NO.	TYPE	
						/	/						
						/	/						
						/	/						
						/	/						
						/	/						
						/	/						
						/	/						
						/	/						
						/	/						
						/	/						
						/	/						
THIS RECORD IS CERTIFIED TRUE AND CORRECT PILOT'S SIGNATURE _____			PAGE TOTAL			/	/						
			PREVIOUS TOTAL			/	/						
			NEW TOTAL			/	/						

	TYPE OF PILOTING TIME					REMARKS AND ENDORSEMENTS
PIC	SIC	AS FLIGHT INSTRUCTOR	DUAL RECEIVED			

20 _____

DATE (MON/DAY)	AIRCRAFT TYPE	AIRCRAFT IDENT	ROUTE OF FLIGHT		FLIGHT NUMBER	TAKE OFF & LANDING			CONDITION OF FLIGHT				FLIGHT SIMULATOR
			FROM	TO		DAY	NIGHT	AUTO LAND	NIGHT	ACTUAL INSTRUMENT	APP R/W NO.	TYPE	
						/	/						
						/	/						
						/	/						
						/	/						
						/	/						
						/	/						
						/	/						
						/	/						
						/	/						
						/	/						
						/	/						
THIS RECORD IS CERTIFIED TRUE AND CORRECT PILOT'S SIGNATURE _____			PAGE TOTAL			/	/						
			PREVIOUS TOTAL			/	/						
			NEW TOTAL			/	/						

	TYPE OF PILOTING TIME					REMARKS AND ENDORSEMENTS
PIC	SIC	AS FLIGHT INSTRUCTOR	DUAL RECEIVED			

20 _____

DATE (MON/DAY)	AIRCRAFT TYPE	AIRCRAFT IDENT	ROUTE OF FLIGHT		FLIGHT NUMBER	TAKE OFF & LANDING			CONDITION OF FLIGHT		APP		FLIGHT SIMULATOR
			FROM	TO		DAY	NIGHT	AUTO LAND	NIGHT	ACTUAL INSTRUMENT	R/W NO.	TYPE	
						/	/						
						/	/						
						/	/						
						/	/						
						/	/						
						/	/						
						/	/						
						/	/						
						/	/						
						/	/						
						/	/						
THIS RECORD IS CERTIFIED TRUE AND CORRECT PILOT'S SIGNATURE _____			PAGE TOTAL			/	/						
			PREVIOUS TOTAL			/	/						
			NEW TOTAL			/	/						

TYPE OF PILOTING TIME						REMARKS AND ENDORSEMENTS
PIC	SIC	AS FLIGHT INSTRUCTOR	DUAL RECEIVED			

20 _____

DATE (MON/DAY)	AIRCRAFT TYPE	AIRCRAFT IDENT	ROUTE OF FLIGHT		FLIGHT NUMBER	TAKE OFF & LANDING			CONDITION OF FLIGHT				FLIGHT SIMULATOR
			FROM	TO		DAY	NIGHT	AUTO LAND	NIGHT	ACTUAL INSTRUMENT	APP R/W NO.	APP TYPE	
						/	/						
						/	/						
						/	/						
						/	/						
						/	/						
						/	/						
						/	/						
						/	/						
						/	/						
						/	/						
						/	/						
						/	/						
THIS RECORD IS CERTIFIED TRUE AND CORRECT PILOT'S SIGNATURE _____			PAGE TOTAL		/	/						
			PREVIOUS TOTAL		/	/						
			NEW TOTAL		/	/						

TYPE OF PILOTING TIME						REMARKS AND ENDORSEMENTS
PIC	SIC	AS FLIGHT INSTRUCTOR	DUAL RECEIVED			

20 _____

DATE (MON/DAY)	AIRCRAFT TYPE	AIRCRAFT IDENT	ROUTE OF FLIGHT		FLIGHT NUMBER	TAKE OFF & LANDING			CONDITION OF FLIGHT				FLIGHT SIMULATOR
			FROM	TO		DAY	NIGHT	AUTO LAND	NIGHT	ACTUAL INSTRUMENT	APP		
											R/W NO.	TYPE	
						/	/						
						/	/						
						/	/						
						/	/						
						/	/						
						/	/						
						/	/						
						/	/						
						/	/						
						/	/						
						/	/						
THIS RECORD IS CERTIFIED TRUE AND CORRECT PILOT'S SIGNATURE _____			PAGE TOTAL			/	/						
			PREVIOUS TOTAL			/	/						
			NEW TOTAL			/	/						

	TYPE OF PILOTING TIME						REMARKS AND ENDORSEMENTS
PIC	SIC	AS FLIGHT INSTRUCTOR	DUAL RECEIVED				

20 _____

DATE (MON/DAY)	AIRCRAFT TYPE	AIRCRAFT IDENT	ROUTE OF FLIGHT		FLIGHT NUMBER	TAKE OFF & LANDING			CONDITION OF FLIGHT				FLIGHT SIMULATOR
			FROM	TO		DAY	NIGHT	AUTO LAND	NIGHT	ACTUAL INSTRUMENT	APP R/W NO.	TYPE	
						/	/						
						/	/						
						/	/						
						/	/						
						/	/						
						/	/						
						/	/						
						/	/						
						/	/						
						/	/						
						/	/						
THIS RECORD IS CERTIFIED TRUE AND CORRECT PILOT'S SIGNATURE _____			PAGE TOTAL		/	/						
			PREVIOUS TOTAL		/	/						
			NEW TOTAL		/	/						

		TYPE OF PILOTING TIME				REMARKS AND ENDORSEMENTS
PIC	SIC	AS FLIGHT INSTRUCTOR	DUAL RECEIVED			

20 _____

DATE (MON/DAY)	AIRCRAFT TYPE	AIRCRAFT IDENT	ROUTE OF FLIGHT		FLIGHT NUMBER	TAKE OFF & LANDING			CONDITION OF FLIGHT				FLIGHT SIMULATOR
			FROM	TO		DAY	NIGHT	AUTO LAND	NIGHT	ACTUAL INSTRUMENT	APP R/W NO.	TYPE	
						/	/						
						/	/						
						/	/						
						/	/						
						/	/						
						/	/						
						/	/						
						/	/						
						/	/						
						/	/						
						/	/						
THIS RECORD IS CERTIFIED TRUE AND CORRECT PILOT'S SIGNATURE _____			PAGE TOTAL			/	/						
			PREVIOUS TOTAL			/	/						
			NEW TOTAL			/	/						

TYPE OF PILOTING TIME						REMARKS AND ENDORSEMENTS
PIC	SIC	AS FLIGHT INSTRUCTOR	DUAL RECEIVED			

20 _____

DATE (MON/DAY)	AIRCRAFT TYPE	AIRCRAFT IDENT	ROUTE OF FLIGHT		FLIGHT NUMBER	TAKE OFF & LANDING			CONDITION OF FLIGHT				FLIGHT SIMULATOR
			FROM	TO		DAY	NIGHT	AUTO LAND	NIGHT	ACTUAL INSTRUMENT	APP		
											R/W NO.	TYPE	
						/	/						
						/	/						
						/	/						
						/	/						
						/	/						
						/	/						
						/	/						
						/	/						
						/	/						
						/	/						
						/	/						
THIS RECORD IS CERTIFIED TRUE AND CORRECT PILOT'S SIGNATURE _____			PAGE TOTAL			/	/						
			PREVIOUS TOTAL			/	/						
			NEW TOTAL			/	/						

	TYPE OF PILOTING TIME					REMARKS AND ENDORSEMENTS
PIC	SIC	AS FLIGHT INSTRUCTOR	DUAL RECEIVED			

20 _____

DATE (MON/DAY)	AIRCRAFT TYPE	AIRCRAFT IDENT	ROUTE OF FLIGHT		FLIGHT NUMBER	TAKE OFF & LANDING			CONDITION OF FLIGHT				FLIGHT SIMULATOR
			FROM	TO		DAY	NIGHT	AUTO LAND	NIGHT	ACTUAL INSTRUMENT	APP R/W NO.	TYPE	
						/	/						
						/	/						
						/	/						
						/	/						
						/	/						
						/	/						
						/	/						
						/	/						
						/	/						
						/	/						
						/	/						

THIS RECORD IS CERTIFIED TRUE AND CORRECT
PILOT'S SIGNATURE _____

	PAGE TOTAL	/	/					
	PREVIOUS TOTAL	/	/					
	NEW TOTAL	/	/					

TYPE OF PILOTING TIME						REMARKS AND ENDORSEMENTS
PIC	SIC	AS FLIGHT INSTRUCTOR	DUAL RECEIVED			

20 _____

DATE (MON/DAY)	AIRCRAFT TYPE	AIRCRAFT IDENT	ROUTE OF FLIGHT		FLIGHT NUMBER	TAKE OFF & LANDING			CONDITION OF FLIGHT				FLIGHT SIMULATOR
			FROM	TO		DAY	NIGHT	AUTO LAND	NIGHT	ACTUAL INSTRUMENT	APP		
											R/W NO.	TYPE	
						/	/						
						/	/						
						/	/						
						/	/						
						/	/						
						/	/						
						/	/						
						/	/						
						/	/						
						/	/						
						/	/						
THIS RECORD IS CERTIFIED TRUE AND CORRECT PILOT'S SIGNATURE _____			PAGE TOTAL			/	/						
			PREVIOUS TOTAL			/	/						
			NEW TOTAL			/	/						

TYPE OF PILOTING TIME						REMARKS AND ENDORSEMENTS
PIC	SIC	AS FLIGHT INSTRUCTOR	DUAL RECEIVED			

20 _____

DATE (MON/DAY)	AIRCRAFT TYPE	AIRCRAFT IDENT	ROUTE OF FLIGHT		FLIGHT NUMBER	TAKE OFF & LANDING			CONDITION OF FLIGHT				FLIGHT SIMULATOR
			FROM	TO		DAY	NIGHT	AUTO LAND	NIGHT	ACTUAL INSTRUMENT	APP R/W NO.	TYPE	
						/	/						
						/	/						
						/	/						
						/	/						
						/	/						
						/	/						
						/	/						
						/	/						
						/	/						
						/	/						
						/	/						
THIS RECORD IS CERTIFIED TRUE AND CORRECT PILOT'S SIGNATURE _____			PAGE TOTAL			/	/						
			PREVIOUS TOTAL			/	/						
			NEW TOTAL			/	/						

TYPE OF PILOTING TIME						REMARKS AND ENDORSEMENTS
PIC	SIC	AS FLIGHT INSTRUCTOR	DUAL RECEIVED			

20 _____

DATE (MON/DAY)	AIRCRAFT TYPE	AIRCRAFT IDENT	ROUTE OF FLIGHT		FLIGHT NUMBER	TAKE OFF & LANDING			CONDITION OF FLIGHT		APP		FLIGHT SIMULATOR
			FROM	TO		DAY	NIGHT	AUTO LAND	NIGHT	ACTUAL INSTRUMENT	R/W NO.	TYPE	
						/	/						
						/	/						
						/	/						
						/	/						
						/	/						
						/	/						
						/	/						
						/	/						
						/	/						
						/	/						
						/	/						
THIS RECORD IS CERTIFIED TRUE AND CORRECT PILOT'S SIGNATURE _____			PAGE TOTAL			/	/						
			PREVIOUS TOTAL			/	/						
			NEW TOTAL			/	/						

TYPE OF PILOTING TIME						REMARKS AND ENDORSEMENTS
PIC	SIC	AS FLIGHT INSTRUCTOR	DUAL RECEIVED			

20 _____

DATE (MON/DAY)	AIRCRAFT TYPE	AIRCRAFT IDENT	ROUTE OF FLIGHT		FLIGHT NUMBER	TAKE OFF & LANDING			CONDITION OF FLIGHT				FLIGHT SIMULATOR
			FROM	TO		DAY	NIGHT	AUTO LAND	NIGHT	ACTUAL INSTRUMENT	APP R/W NO.	TYPE	
						/	/						
						/	/						
						/	/						
						/	/						
						/	/						
						/	/						
						/	/						
						/	/						
						/	/						
						/	/						
						/	/						
THIS RECORD IS CERTIFIED TRUE AND CORRECT PILOT'S SIGNATURE _____			PAGE TOTAL			/	/						
			PREVIOUS TOTAL			/	/						
			NEW TOTAL			/	/						

TYPE OF PILOTING TIME						REMARKS AND ENDORSEMENTS
PIC	SIC	AS FLIGHT INSTRUCTOR	DUAL RECEIVED			

20 _____

DATE (MON/DAY)	AIRCRAFT TYPE	AIRCRAFT IDENT	ROUTE OF FLIGHT		FLIGHT NUMBER	TAKE OFF & LANDING			CONDITION OF FLIGHT				FLIGHT SIMULATOR
			FROM	TO		DAY	NIGHT	AUTO LAND	NIGHT	ACTUAL INSTRUMENT	APP R/W NO.	TYPE	
						/	/						
						/	/						
						/	/						
						/	/						
						/	/						
						/	/						
						/	/						
						/	/						
						/	/						
						/	/						
						/	/						
THIS RECORD IS CERTIFIED TRUE AND CORRECT PILOT'S SIGNATURE _____			PAGE TOTAL			/	/						
			PREVIOUS TOTAL			/	/						
			NEW TOTAL			/	/						

TYPE OF PILOTING TIME						REMARKS AND ENDORSEMENTS
PIC	SIC	AS FLIGHT INSTRUCTOR	DUAL RECEIVED			

20 _____

DATE (MON/DAY)	AIRCRAFT TYPE	AIRCRAFT IDENT	ROUTE OF FLIGHT		FLIGHT NUMBER	TAKE OFF & LANDING			CONDITION OF FLIGHT				FLIGHT SIMULATOR
			FROM	TO		DAY	NIGHT	AUTO LAND	NIGHT	ACTUAL INSTRUMENT	APP R/W NO.	TYPE	
						/	/						
						/	/						
						/	/						
						/	/						
						/	/						
						/	/						
						/	/						
						/	/						
						/	/						
						/	/						
						/	/						
THIS RECORD IS CERTIFIED TRUE AND CORRECT PILOT'S SIGNATURE _____			PAGE TOTAL			/	/						
			PREVIOUS TOTAL			/	/						
			NEW TOTAL			/	/						

TYPE OF PILOTING TIME						REMARKS AND ENDORSEMENTS
PIC	SIC	AS FLIGHT INSTRUCTOR	DUAL RECEIVED			

20 _____

DATE (MON/DAY)	AIRCRAFT TYPE	AIRCRAFT IDENT	ROUTE OF FLIGHT		FLIGHT NUMBER	TAKE OFF & LANDING			CONDITION OF FLIGHT				FLIGHT SIMULATOR
			FROM	TO		DAY	NIGHT	AUTO LAND	NIGHT	ACTUAL INSTRUMENT	APP R/W NO.	TYPE	
						/	/						
						/	/						
						/	/						
						/	/						
						/	/						
						/	/						
						/	/						
						/	/						
						/	/						
						/	/						
						/	/						
THIS RECORD IS CERTIFIED TRUE AND CORRECT PILOT'S SIGNATURE _____			PAGE TOTAL			/	/						
			PREVIOUS TOTAL			/	/						
			NEW TOTAL			/	/						

TYPE OF PILOTING TIME						REMARKS AND ENDORSEMENTS
PIC	SIC	AS FLIGHT INSTRUCTOR	DUAL RECEIVED			

20 _____

DATE (MON/DAY)	AIRCRAFT TYPE	AIRCRAFT IDENT	ROUTE OF FLIGHT		FLIGHT NUMBER	TAKE OFF & LANDING			CONDITION OF FLIGHT		APP		FLIGHT SIMULATOR
			FROM	TO		DAY	NIGHT	AUTO LAND	NIGHT	ACTUAL INSTRUMENT	R/W NO.	TYPE	
						/	/						
						/	/						
						/	/						
						/	/						
						/	/						
						/	/						
						/	/						
						/	/						
						/	/						
						/	/						
						/	/						
THIS RECORD IS CERTIFIED TRUE AND CORRECT PILOT'S SIGNATURE _____			PAGE TOTAL			/	/						
			PREVIOUS TOTAL			/	/						
			NEW TOTAL			/	/						

TYPE OF PILOTING TIME						REMARKS AND ENDORSEMENTS
PIC	SIC	AS FLIGHT INSTRUCTOR	DUAL RECEIVED			

20 _____

DATE (MON/DAY)	AIRCRAFT TYPE	AIRCRAFT IDENT	ROUTE OF FLIGHT		FLIGHT NUMBER	TAKE OFF & LANDING			CONDITION OF FLIGHT				FLIGHT SIMULATOR
			FROM	TO		DAY	NIGHT	AUTO LAND	NIGHT	ACTUAL INSTRUMENT	APP R/W NO.	TYPE	
						/	/						
						/	/						
						/	/						
						/	/						
						/	/						
						/	/						
						/	/						
						/	/						
						/	/						
						/	/						
						/	/						
THIS RECORD IS CERTIFIED TRUE AND CORRECT PILOT'S SIGNATURE _____			PAGE TOTAL			/	/						
			PREVIOUS TOTAL			/	/						
			NEW TOTAL			/	/						

	TYPE OF PILOTING TIME					REMARKS AND ENDORSEMENTS
PIC	SIC	AS FLIGHT INSTRUCTOR	DUAL RECEIVED			

20 _____

DATE (MON/DAY)	AIRCRAFT TYPE	AIRCRAFT IDENT	ROUTE OF FLIGHT		FLIGHT NUMBER	TAKE OFF & LANDING			CONDITION OF FLIGHT				FLIGHT SIMULATOR
			FROM	TO		DAY	NIGHT	AUTO LAND	NIGHT	ACTUAL INSTRUMENT	APP		
											R/W NO.	TYPE	
						/	/						
						/	/						
						/	/						
						/	/						
						/	/						
						/	/						
						/	/						
						/	/						
						/	/						
						/	/						
						/	/						
THIS RECORD IS CERTIFIED TRUE AND CORRECT PILOT'S SIGNATURE _____			PAGE TOTAL			/	/						
			PREVIOUS TOTAL			/	/						
			NEW TOTAL			/	/						

TYPE OF PILOTING TIME						REMARKS AND ENDORSEMENTS
PIC	SIC	AS FLIGHT INSTRUCTOR	DUAL RECEIVED			

20 _____

DATE (MON/DAY)	AIRCRAFT TYPE	AIRCRAFT IDENT	ROUTE OF FLIGHT		FLIGHT NUMBER	TAKE OFF & LANDING			CONDITION OF FLIGHT		APP		FLIGHT SIMULATOR
			FROM	TO		DAY	NIGHT	AUTO LAND	NIGHT	ACTUAL INSTRUMENT	R/W NO.	TYPE	
						/	/						
						/	/						
						/	/						
						/	/						
						/	/						
						/	/						
						/	/						
						/	/						
						/	/						
						/	/						
						/	/						
THIS RECORD IS CERTIFIED TRUE AND CORRECT PILOT'S SIGNATURE _____			PAGE TOTAL			/	/						
			PREVIOUS TOTAL			/	/						
			NEW TOTAL			/	/						

TYPE OF PILOTING TIME							REMARKS AND ENDORSEMENTS
PIC	SIC	AS FLIGHT INSTRUCTOR	DUAL RECEIVED			

20 _____

DATE (MON/DAY)	AIRCRAFT TYPE	AIRCRAFT IDENT	ROUTE OF FLIGHT		FLIGHT NUMBER	TAKE OFF & LANDING			CONDITION OF FLIGHT		APP		FLIGHT SIMULATOR
			FROM	TO		DAY	NIGHT	AUTO LAND	NIGHT	ACTUAL INSTRUMENT	R/W NO.	TYPE	
						/	/						
						/	/						
						/	/						
						/	/						
						/	/						
						/	/						
						/	/						
						/	/						
						/	/						
						/	/						
						/	/						

THIS RECORD IS CERTIFIED TRUE AND CORRECT PILOT'S SIGNATURE _____

	PAGE TOTAL	/	/						
	PREVIOUS TOTAL	/	/						
	NEW TOTAL	/	/						

	TYPE OF PILOTING TIME						REMARKS AND ENDORSEMENTS
PIC	SIC	AS FLIGHT INSTRUCTOR	DUAL RECEIVED				

20 _____

DATE (MON/DAY)	AIRCRAFT TYPE	AIRCRAFT IDENT	ROUTE OF FLIGHT		FLIGHT NUMBER	TAKE OFF & LANDING			CONDITION OF FLIGHT				FLIGHT SIMULATOR
			FROM	TO		DAY	NIGHT	AUTO LAND	NIGHT	ACTUAL INSTRUMENT	APP R/W NO.	TYPE	
						/	/						
						/	/						
						/	/						
						/	/						
						/	/						
						/	/						
						/	/						
						/	/						
						/	/						
						/	/						
						/	/						
THIS RECORD IS CERTIFIED TRUE AND CORRECT PILOT'S SIGNATURE _____			PAGE TOTAL			/	/						
			PREVIOUS TOTAL			/	/						
			NEW TOTAL			/	/						

	TYPE OF PILOTING TIME					REMARKS AND ENDORSEMENTS
PIC	SIC	AS FLIGHT INSTRUCTOR	DUAL RECEIVED			

20 _____

DATE (MON/DAY)	AIRCRAFT TYPE	AIRCRAFT IDENT	ROUTE OF FLIGHT		FLIGHT NUMBER	TAKE OFF & LANDING			CONDITION OF FLIGHT				FLIGHT SIMULATOR
			FROM	TO		DAY	NIGHT	AUTO LAND	NIGHT	ACTUAL INSTRUMENT	APP R/W NO.	TYPE	
						/	/						
						/	/						
						/	/						
						/	/						
						/	/						
						/	/						
						/	/						
						/	/						
						/	/						
						/	/						
						/	/						
THIS RECORD IS CERTIFIED TRUE AND CORRECT PILOT'S SIGNATURE _____			PAGE TOTAL			/	/						
			PREVIOUS TOTAL			/	/						
			NEW TOTAL			/	/						

	TYPE OF PILOTING TIME					REMARKS AND ENDORSEMENTS
PIC	SIC	AS FLIGHT INSTRUCTOR	DUAL RECEIVED			

20 _____

DATE (MON/DAY)	AIRCRAFT TYPE	AIRCRAFT IDENT	ROUTE OF FLIGHT		FLIGHT NUMBER	TAKE OFF & LANDING			CONDITION OF FLIGHT				FLIGHT SIMULATOR
			FROM	TO		DAY	NIGHT	AUTO LAND	NIGHT	ACTUAL INSTRUMENT	APP R/W NO.	TYPE	
						/	/						
						/	/						
						/	/						
						/	/						
						/	/						
						/	/						
						/	/						
						/	/						
						/	/						
						/	/						
						/	/						
THIS RECORD IS CERTIFIED TRUE AND CORRECT PILOT'S SIGNATURE _____			PAGE TOTAL			/	/						
			PREVIOUS TOTAL			/	/						
			NEW TOTAL			/	/						

TYPE OF PILOTING TIME						REMARKS AND ENDORSEMENTS
PIC	SIC	AS FLIGHT INSTRUCTOR	DUAL RECEIVED			

20 _____

DATE (MON/DAY)	AIRCRAFT TYPE	AIRCRAFT IDENT	ROUTE OF FLIGHT		FLIGHT NUMBER	TAKE OFF & LANDING			CONDITION OF FLIGHT		APP		FLIGHT SIMULATOR
			FROM	TO		DAY	NIGHT	AUTO LAND	NIGHT	ACTUAL INSTRUMENT	R/W NO.	TYPE	
						/	/						
						/	/						
						/	/						
						/	/						
						/	/						
						/	/						
						/	/						
						/	/						
						/	/						
						/	/						
						/	/						
THIS RECORD IS CERTIFIED TRUE AND CORRECT PILOT'S SIGNATURE _____			PAGE TOTAL			/	/						
			PREVIOUS TOTAL			/	/						
			NEW TOTAL			/	/						

		TYPE OF PILOTING TIME				REMARKS AND ENDORSEMENTS
PIC	SIC	AS FLIGHT INSTRUCTOR	DUAL RECEIVED			

20 _____

DATE (MON/DAY)	AIRCRAFT TYPE	AIRCRAFT IDENT	ROUTE OF FLIGHT		FLIGHT NUMBER	TAKE OFF & LANDING			CONDITION OF FLIGHT				FLIGHT SIMULATOR
			FROM	TO		DAY	NIGHT	AUTO LAND	NIGHT	ACTUAL INSTRUMENT	APP R/W NO.	TYPE	
						/	/						
						/	/						
						/	/						
						/	/						
						/	/						
						/	/						
						/	/						
						/	/						
						/	/						
						/	/						
						/	/						
THIS RECORD IS CERTIFIED TRUE AND CORRECT PILOT'S SIGNATURE _____			PAGE TOTAL			/	/						
			PREVIOUS TOTAL			/	/						
			NEW TOTAL			/	/						

	TYPE OF PILOTING TIME					REMARKS AND ENDORSEMENTS
PIC	SIC	AS FLIGHT INSTRUCTOR	DUAL RECEIVED			

20 _____

DATE (MON/DAY)	AIRCRAFT TYPE	AIRCRAFT IDENT	ROUTE OF FLIGHT		FLIGHT NUMBER	TAKE OFF & LANDING			CONDITION OF FLIGHT				FLIGHT SIMULATOR
			FROM	TO		DAY	NIGHT	AUTO LAND	NIGHT	ACTUAL INSTRUMENT	APP R/W NO.	TYPE	
						/	/						
						/	/						
						/	/						
						/	/						
						/	/						
						/	/						
						/	/						
						/	/						
						/	/						
						/	/						
						/	/						
THIS RECORD IS CERTIFIED TRUE AND CORRECT PILOT'S SIGNATURE _____			PAGE TOTAL			/	/						
			PREVIOUS TOTAL			/	/						
			NEW TOTAL			/	/						

TYPE OF PILOTING TIME							REMARKS AND ENDORSEMENTS
PIC	SIC	AS FLIGHT INSTRUCTOR	DUAL RECEIVED			

20 _____

DATE (MON/DAY)	AIRCRAFT TYPE	AIRCRAFT IDENT	ROUTE OF FLIGHT		FLIGHT NUMBER	TAKE OFF & LANDING			CONDITION OF FLIGHT		APP		FLIGHT SIMULATOR
			FROM	TO		DAY	NIGHT	AUTO LAND	NIGHT	ACTUAL INSTRUMENT	R/W NO.	TYPE	
						/	/						
						/	/						
						/	/						
						/	/						
						/	/						
						/	/						
						/	/						
						/	/						
						/	/						
						/	/						
						/	/						
THIS RECORD IS CERTIFIED TRUE AND CORRECT PILOT'S SIGNATURE _____			PAGE TOTAL			/	/						
			PREVIOUS TOTAL			/	/						
			NEW TOTAL			/	/						

TYPE OF PILOTING TIME						REMARKS AND ENDORSEMENTS
PIC	SIC	AS FLIGHT INSTRUCTOR	DUAL RECEIVED			

20 _____

DATE (MON/DAY)	AIRCRAFT TYPE	AIRCRAFT IDENT	ROUTE OF FLIGHT		FLIGHT NUMBER	TAKE OFF & LANDING			CONDITION OF FLIGHT				FLIGHT SIMULATOR
			FROM	TO		DAY	NIGHT	AUTO LAND	NIGHT	ACTUAL INSTRUMENT	APP R/W NO.	TYPE	
						/	/						
						/	/						
						/	/						
						/	/						
						/	/						
						/	/						
						/	/						
						/	/						
						/	/						
						/	/						
						/	/						
THIS RECORD IS CERTIFIED TRUE AND CORRECT PILOT'S SIGNATURE _____			PAGE TOTAL			/	/						
			PREVIOUS TOTAL			/	/						
			NEW TOTAL			/	/						

	TYPE OF PILOTING TIME					REMARKS AND ENDORSEMENTS
PIC	SIC	AS FLIGHT INSTRUCTOR	DUAL RECEIVED			

20 _____

DATE (MON/DAY)	AIRCRAFT TYPE	AIRCRAFT IDENT	ROUTE OF FLIGHT		FLIGHT NUMBER	TAKE OFF & LANDING			CONDITION OF FLIGHT				FLIGHT SIMULATOR
			FROM	TO		DAY	NIGHT	AUTO LAND	NIGHT	ACTUAL INSTRUMENT	APP R/W NO.	APP TYPE	
						/	/						
						/	/						
						/	/						
						/	/						
						/	/						
						/	/						
						/	/						
						/	/						
						/	/						
						/	/						
						/	/						
THIS RECORD IS CERTIFIED TRUE AND CORRECT PILOT'S SIGNATURE _____			PAGE TOTAL			/	/						
			PREVIOUS TOTAL			/	/						
			NEW TOTAL			/	/						

	TYPE OF PILOTING TIME					REMARKS AND ENDORSEMENTS
PIC	SIC	AS FLIGHT INSTRUCTOR	DUAL RECEIVED			

20 _____

DATE (MON/DAY)	AIRCRAFT TYPE	AIRCRAFT IDENT	ROUTE OF FLIGHT		FLIGHT NUMBER	TAKE OFF & LANDING			CONDITION OF FLIGHT		APP		FLIGHT SIMULATOR
			FROM	TO		DAY	NIGHT	AUTO LAND	NIGHT	ACTUAL INSTRUMENT	R/W NO.	TYPE	
						/	/						
						/	/						
						/	/						
						/	/						
						/	/						
						/	/						
						/	/						
						/	/						
						/	/						
						/	/						
						/	/						
THIS RECORD IS CERTIFIED TRUE AND CORRECT PILOT'S SIGNATURE _____			PAGE TOTAL			/	/						
			PREVIOUS TOTAL			/	/						
			NEW TOTAL			/	/						

		TYPE OF PILOTING TIME				REMARKS AND ENDORSEMENTS
PIC	SIC	AS FLIGHT INSTRUCTOR	DUAL RECEIVED			

20 _____

DATE (MON/DAY)	AIRCRAFT TYPE	AIRCRAFT IDENT	ROUTE OF FLIGHT		FLIGHT NUMBER	TAKE OFF & LANDING			CONDITION OF FLIGHT		APP		FLIGHT SIMULATOR
			FROM	TO		DAY	NIGHT	AUTO LAND	NIGHT	ACTUAL INSTRUMENT	R/W NO.	TYPE	
						/	/						
						/	/						
						/	/						
						/	/						
						/	/						
						/	/						
						/	/						
						/	/						
						/	/						
						/	/						
						/	/						
THIS RECORD IS CERTIFIED TRUE AND CORRECT PILOT'S SIGNATURE _____			PAGE TOTAL			/	/						
			PREVIOUS TOTAL			/	/						
			NEW TOTAL			/	/						

		TYPE OF PILOTING TIME				REMARKS AND ENDORSEMENTS
PIC	SIC	AS FLIGHT INSTRUCTOR	DUAL RECEIVED			

20_____

| DATE (MON/DAY) | AIRCRAFT TYPE | AIRCRAFT IDENT | ROUTE OF FLIGHT | | FLIGHT NUMBER | TAKE OFF & LANDING | | | CONDITION OF FLIGHT | | | | FLIGHT SIMULATOR |
| | | | FROM | TO | | DAY | NIGHT | AUTO LAND | NIGHT | ACTUAL INSTRUMENT | APP | | |
											R/W NO.	TYPE	
						/	/						
						/	/						
						/	/						
						/	/						
						/	/						
						/	/						
						/	/						
						/	/						
						/	/						
						/	/						
						/	/						
THIS RECORD IS CERTIFIED TRUE AND CORRECT PILOT'S SIGNATURE _____			PAGE TOTAL			/	/						
			PREVIOUS TOTAL			/	/						
			NEW TOTAL			/	/						

	TYPE OF PILOTING TIME					REMARKS AND ENDORSEMENTS
PIC	SIC	AS FLIGHT INSTRUCTOR	DUAL RECEIVED			

20 _____

DATE (MON/DAY)	AIRCRAFT TYPE	AIRCRAFT IDENT	ROUTE OF FLIGHT		FLIGHT NUMBER	TAKE OFF & LANDING			CONDITION OF FLIGHT		APP		FLIGHT SIMULATOR
			FROM	TO		DAY	NIGHT	AUTO LAND	NIGHT	ACTUAL INSTRUMENT	R/W NO.	TYPE	
						/	/						
						/	/						
						/	/						
						/	/						
						/	/						
						/	/						
						/	/						
						/	/						
						/	/						
						/	/						
						/	/						
THIS RECORD IS CERTIFIED TRUE AND CORRECT PILOT'S SIGNATURE _____			PAGE TOTAL			/	/						
			PREVIOUS TOTAL			/	/						
			NEW TOTAL			/	/						

TYPE OF PILOTING TIME							REMARKS AND ENDORSEMENTS
PIC	SIC	AS FLIGHT INSTRUCTOR	DUAL RECEIVED			

20 _____

DATE (MON/DAY)	AIRCRAFT TYPE	AIRCRAFT IDENT	ROUTE OF FLIGHT		FLIGHT NUMBER	TAKE OFF & LANDING			CONDITION OF FLIGHT				FLIGHT SIMULATOR
			FROM	TO		DAY	NIGHT	AUTO LAND	NIGHT	ACTUAL INSTRUMENT	APP		
											R/W NO.	TYPE	
						/	/						
						/	/						
						/	/						
						/	/						
						/	/						
						/	/						
						/	/						
						/	/						
						/	/						
						/	/						
						/	/						
THIS RECORD IS CERTIFIED TRUE AND CORRECT PILOT'S SIGNATURE _____			PAGE TOTAL		/	/						
			PREVIOUS TOTAL		/	/						
			NEW TOTAL		/	/						

PIC	SIC	AS FLIGHT INSTRUCTOR	DUAL RECEIVED			REMARKS AND ENDORSEMENTS

TYPE OF PILOTING TIME

20 _____

DATE (MON/DAY)	AIRCRAFT TYPE	AIRCRAFT IDENT	ROUTE OF FLIGHT		FLIGHT NUMBER	TAKE OFF & LANDING			CONDITION OF FLIGHT		APP		FLIGHT SIMULATOR
			FROM	TO		DAY	NIGHT	AUTO LAND	NIGHT	ACTUAL INSTRUMENT	R/W NO.	TYPE	
						/	/						
						/	/						
						/	/						
						/	/						
						/	/						
						/	/						
						/	/						
						/	/						
						/	/						
						/	/						
						/	/						
THIS RECORD IS CERTIFIED TRUE AND CORRECT PILOT'S SIGNATURE _____			PAGE TOTAL			/	/						
			PREVIOUS TOTAL			/	/ .						
			NEW TOTAL			/	/						

	TYPE OF PILOTING TIME					REMARKS AND ENDORSEMENTS
PIC	SIC	AS FLIGHT INSTRUCTOR	DUAL RECEIVED			

20 _____

DATE (MON/DAY)	AIRCRAFT TYPE	AIRCRAFT IDENT	ROUTE OF FLIGHT		FLIGHT NUMBER	TAKE OFF & LANDING			CONDITION OF FLIGHT				FLIGHT SIMULATOR
			FROM	TO		DAY	NIGHT	AUTO LAND	NIGHT	ACTUAL INSTRUMENT	APP R/W NO.	TYPE	
						/	/						
						/	/						
						/	/						
						/	/						
						/	/						
						/	/						
						/	/						
						/	/						
						/	/						
						/	/						
						/	/						
THIS RECORD IS CERTIFIED TRUE AND CORRECT PILOT'S SIGNATURE _____			PAGE TOTAL			/	/						
			PREVIOUS TOTAL			/	/						
			NEW TOTAL			/	/						

TYPE OF PILOTING TIME						REMARKS AND ENDORSEMENTS
PIC	SIC	AS FLIGHT INSTRUCTOR	DUAL RECEIVED			

20 _____

DATE (MON/DAY)	AIRCRAFT TYPE	AIRCRAFT IDENT	ROUTE OF FLIGHT		FLIGHT NUMBER	TAKE OFF & LANDING			CONDITION OF FLIGHT				FLIGHT SIMULATOR
			FROM	TO		DAY	NIGHT	AUTO LAND	NIGHT	ACTUAL INSTRUMENT	APP R/W NO.	TYPE	
						/	/						
						/	/						
						/	/						
						/	/						
						/	/						
						/	/						
						/	/						
						/	/						
						/	/						
						/	/						
						/	/						
THIS RECORD IS CERTIFIED TRUE AND CORRECT PILOT'S SIGNATURE _____			PAGE TOTAL			/	/						
			PREVIOUS TOTAL			/	/						
			NEW TOTAL			/	/						

TYPE OF PILOTING TIME						REMARKS AND ENDORSEMENTS
PIC	SIC	AS FLIGHT INSTRUCTOR	DUAL RECEIVED			

20 _____

| DATE (MON/DAY) | AIRCRAFT TYPE | AIRCRAFT IDENT | ROUTE OF FLIGHT || FLIGHT NUMBER | TAKE OFF & LANDING ||| CONDITION OF FLIGHT |||| FLIGHT SIMULATOR |
			FROM	TO		DAY	NIGHT	AUTO LAND	NIGHT	ACTUAL INSTRUMENT	APP R/W NO.	APP TYPE	
						/	/						
						/	/						
						/	/						
						/	/						
						/	/						
						/	/						
						/	/						
						/	/						
						/	/						
						/	/						
						/	/						
THIS RECORD IS CERTIFIED TRUE AND CORRECT PILOT'S SIGNATURE _____			PAGE TOTAL			/	/						
			PREVIOUS TOTAL			/	/						
			NEW TOTAL			/	/						

TYPE OF PILOTING TIME						REMARKS AND ENDORSEMENTS
PIC	SIC	AS FLIGHT INSTRUCTOR	DUAL RECEIVED			

20 _____

DATE (MON/DAY)	AIRCRAFT TYPE	AIRCRAFT IDENT	ROUTE OF FLIGHT		FLIGHT NUMBER	TAKE OFF & LANDING			CONDITION OF FLIGHT		APP		FLIGHT SIMULATOR
			FROM	TO		DAY	NIGHT	AUTO LAND	NIGHT	ACTUAL INSTRUMENT	R/W NO.	TYPE	
						/	/						
						/	/						
						/	/						
						/	/						
						/	/						
						/	/						
						/	/						
						/	/						
						/	/						
						/	/						
						/	/						
THIS RECORD IS CERTIFIED TRUE AND CORRECT PILOT'S SIGNATURE _____			PAGE TOTAL			/	/						
			PREVIOUS TOTAL			/	/						
			NEW TOTAL			/	/						

	TYPE OF PILOTING TIME						REMARKS AND ENDORSEMENTS
PIC	SIC	AS FLIGHT INSTRUCTOR	DUAL RECEIVED				

20 _____

DATE (MON/DAY)	AIRCRAFT TYPE	AIRCRAFT IDENT	ROUTE OF FLIGHT		FLIGHT NUMBER	TAKE OFF & LANDING			CONDITION OF FLIGHT		APP		FLIGHT SIMULATOR
			FROM	TO		DAY	NIGHT	AUTO LAND	NIGHT	ACTUAL INSTRUMENT	R/W NO.	TYPE	
						/	/						
						/	/						
						/	/						
						/	/						
						/	/						
						/	/						
						/	/						
						/	/						
						/	/						
						/	/						
						/	/						
THIS RECORD IS CERTIFIED TRUE AND CORRECT PILOT'S SIGNATURE _____			PAGE TOTAL			/	/						
			PREVIOUS TOTAL			/	/						
			NEW TOTAL			/	/						

	TYPE OF PILOTING TIME					REMARKS AND ENDORSEMENTS
PIC	SIC	AS FLIGHT INSTRUCTOR	DUAL RECEIVED			

20_____

DATE (MON/DAY)	AIRCRAFT TYPE	AIRCRAFT IDENT	ROUTE OF FLIGHT		FLIGHT NUMBER	TAKE OFF & LANDING			CONDITION OF FLIGHT		APP		FLIGHT SIMULATOR
			FROM	TO		DAY	NIGHT	AUTO LAND	NIGHT	ACTUAL INSTRUMENT	R/W NO.	TYPE	
						/	/						
						/	/						
						/	/						
						/	/						
						/	/						
						/	/						
						/	/						
						/	/						
						/	/						
						/	/						
						/	/						
THIS RECORD IS CERTIFIED TRUE AND CORRECT PILOT'S SIGNATURE _____			PAGE TOTAL			/	/						
			PREVIOUS TOTAL			/	/						
			NEW TOTAL			/	/						

TYPE OF PILOTING TIME						REMARKS AND ENDORSEMENTS
PIC	SIC	AS FLIGHT INSTRUCTOR	DUAL RECEIVED			

20 _____

DATE (MON/DAY)	AIRCRAFT TYPE	AIRCRAFT IDENT	ROUTE OF FLIGHT		FLIGHT NUMBER	TAKE OFF & LANDING			CONDITION OF FLIGHT				FLIGHT SIMULATOR
			FROM	TO		DAY	NIGHT	AUTO LAND	NIGHT	ACTUAL INSTRUMENT	APP		
											R/W NO.	TYPE	
						/	/						
						/	/						
						/	/						
						/	/						
						/	/						
						/	/						
						/	/						
						/	/						
						/	/						
						/	/						
						/	/						
THIS RECORD IS CERTIFIED TRUE AND CORRECT PILOT'S SIGNATURE _____			PAGE TOTAL			/	/						
			PREVIOUS TOTAL			/	/						
			NEW TOTAL			/	/						

TYPE OF PILOTING TIME						REMARKS AND ENDORSEMENTS
PIC	SIC	AS FLIGHT INSTRUCTOR	DUAL RECEIVED			

발행 2023년 1월 10일 1판3쇄 발행 **발행처** 도서출판세화 **펴낸이** 박 용
등록일자 1978년 12월 26일 제 1-338호 **주소** 경기도 파주시 회동길 325-22(서패동 469-2)
편집부 (031)955_9333 영업부 (02)719_3142, (031)955_9331~2 **팩스** (02)719_3146, (031)955_9334
웹사이트 www.sehwapub.co.kr

정가 15,000원 ISBN 978-89-317-0887-5 13550